CRYPTO & BEYOND:

A Beginner's Guide to Understanding the World of Cryptocurrencies

Justin M. Hill

Index

Chapter 1

The Rise of Cryptocurrencies:

A Brief Overview of their Origins and Evolution

In recent years, cryptocurrencies have become increasingly popular and have gained significant attention from investors, businesses, and governments around the world. These digital currencies have revolutionized the way we think about money and have the potential to change the way we conduct financial transactions. In this essay, we will explore the origins and evolution of cryptocurrencies, from the creation of Bitcoin to the current state of the market.

Origins of Cryptocurrencies

The idea of a digital currency is not new, and many attempts were made in the past to create a secure and decentralized system of money. However, it was not until the creation of Bitcoin in 2009 that the concept of cryptocurrencies truly took off. Bitcoin was invented by an unknown person or group using the pseudonym Satoshi Nakamoto. The goal was to create a decentralized, digital currency that would allow for secure and private transactions without the need for intermediaries, such as banks or governments.

Bitcoin was created using a new technology called blockchain. The blockchain is a decentralized digital ledger that records every transaction made using the currency. Each block in the blockchain contains a unique cryptographic code, and once a block is added to the chain, it cannot be altered. This makes the blockchain a tamper-proof and secure system for storing and transferring digital assets.

Evolution of Cryptocurrencies

After the creation of Bitcoin, other cryptocurrencies quickly followed, and today, there are thousands of different digital currencies in existence. Each cryptocurrency has its own unique features, but they all share the same fundamental principles of decentralization, security, and privacy.

One of the main benefits of cryptocurrencies is their decentralized nature. Unlike traditional currencies that are controlled by governments and banks, cryptocurrencies are not subject to central control. This means that they are not affected by inflation, interest rates, or other economic factors that can impact traditional currencies. Instead, the value of cryptocurrencies is determined by the market, based on supply and demand.

Another benefit of cryptocurrencies is their security. The blockchain technology that underpins cryptocurrencies is highly secure and difficult to hack. This makes cryptocurrencies an attractive option for those who want to keep their financial transactions private and secure.

The evolution of cryptocurrencies has been driven by a number of factors, including increased awareness and adoption, improved technology, and changes in regulations. As more people learn about cryptocurrencies and their benefits, demand for these digital currencies continues to grow. This has led to an increase in the number of businesses and organizations that are accepting cryptocurrencies as payment, and many governments are exploring the possibility of creating their own digital currencies.

In addition to Bitcoin, other cryptocurrencies that have gained popularity include Ethereum, Ripple, Litecoin, and Bitcoin Cash. Each of these cryptocurrencies has its own unique features, and some have been specifically designed for certain use cases, such as smart contracts or cross-border payments.

Challenges and Controversies

Despite the many benefits of cryptocurrencies, there are also challenges and controversies associated with them. One of the biggest challenges facing cryptocurrencies is their volatility. The value of cryptocurrencies can fluctuate rapidly, making them a risky investment for those who are not familiar with the market.

Cryptocurrencies have also been associated with a number of controversies, including their use in illegal activities such as money laundering and terrorism financing. Additionally, some critics argue that cryptocurrencies are not backed by any tangible assets and that their value is based solely on speculation.

Regulations and Future of Cryptocurrencies

As cryptocurrencies continue to gain popularity, governments and regulators around the world are grappling with how to regulate this new form of currency. Some countries, such as Japan and Switzerland, have adopted a friendly stance toward cryptocurrencies, while others, such as China and India, have banned cryptocurrencies altogether or placed heavy restrictions on their use.

The lack of global regulations and the differing opinions of governments and regulators on cryptocurrencies have created an uncertain environment for investors and businesses. However, many experts believe that regulations are necessary to ensure the legitimacy and stability of the cryptocurrency market.

In the future, it is likely that cryptocurrencies will continue to evolve and adapt to changing market conditions and technological advancements. Some experts predict that cryptocurrencies could become the dominant form of currency in the digital age, while others believe that they will remain a niche market.

Regardless of the outcome, it is clear that cryptocurrencies have already had a significant impact on the financial industry, and their influence is only expected to grow in the coming years. With their unique features and potential benefits, cryptocurrencies have the power to revolutionize the way we think about money and financial transactions.

In conclusion,

cryptocurrencies have come a long way since the creation of Bitcoin in 2009. The rise of cryptocurrencies has been driven by their decentralized nature, security, and privacy, which have made them an attractive option for investors and businesses. However, the market is not without its challenges and controversies, and regulations are needed to ensure the legitimacy and stability of the market.

Despite the uncertainties and challenges facing cryptocurrencies, they have already had a significant impact on the financial industry, and their influence is expected to grow in the coming years. As the market continues to evolve and adapt, it is important for investors and businesses to stay informed and educated about this new form of currency.

Chapter 2

"Understanding Blockchain Technology: How it Powers the Crypto Revolution"

Blockchain technology is at the heart of the cryptocurrency revolution. It is a decentralized, digital ledger that allows for secure and private transactions without the need for intermediaries, such as banks or governments. In this essay, we will explore the basics of blockchain technology, how it works, and how it powers the cryptocurrency market.

What is Blockchain Technology?

At its most basic level, blockchain technology is a decentralized database that is maintained by a network of computers. Each computer on the network has a copy of the database, which is constantly updated to reflect new transactions. Once a transaction is recorded on the blockchain, it cannot be altered or deleted. This makes the blockchain a secure and tamper-proof system for storing and transferring digital assets.

The blockchain is made up of blocks, which are linked together in chronological order to form a chain. Each block contains a unique cryptographic code, called a hash, that verifies the integrity of the block. Once a block is added to the blockchain, it cannot be changed, which ensures the immutability of the ledger.

How Does Blockchain Technology Work?

The blockchain is a distributed ledger, which means that it is not stored in a central location but is instead spread across a network of computers. Each computer on the network, known as a node, has a copy of the blockchain and participates in the validation and verification of new transactions.

When a new transaction is initiated, it is broadcast to the network of nodes. Each node on the network verifies the transaction to ensure that it is valid and has not already been recorded on the blockchain. Once the transaction is verified, it is added to a new block, which is then broadcast to the network for validation.

Once a block is validated by the network, it is added to the blockchain, and the transaction is considered complete. The transaction is now visible to anyone on the network, and the ledger is updated to reflect the new transaction.

Why is Blockchain Technology Important?

Blockchain technology is important because it provides a secure and decentralized system for storing and transferring digital assets. This is particularly important in the context of the cryptocurrency market, where traditional forms of currency are not suitable.

In the traditional financial system, transactions are often subject to delays, fees, and the risk of fraud. Blockchain technology eliminates many of these risks by providing a system that is secure, transparent, and decentralized.

Additionally, blockchain technology has the potential to disrupt many other industries beyond finance. The immutability and transparency of the blockchain can be used to create secure and tamper-proof systems for voting, supply chain management, and more.

Challenges and Controversies

Despite the many benefits of blockchain technology, there are also challenges and controversies associated with it. One of the biggest challenges facing blockchain technology is scalability. As the number of transactions on the blockchain increases, the size of the database also increases, which can lead to slower transaction times and higher costs.

Additionally, the anonymity of the blockchain has made it attractive to criminals, who have used it for illegal activities such as money laundering and terrorism financing. This has led to concerns from governments and regulators about the potential risks associated with cryptocurrencies and the blockchain.

Regulations and Future of Blockchain Technology

As the blockchain continues to gain popularity and adoption, governments and regulators around the world are grappling with how to regulate this new technology. Some countries, such as Japan and Switzerland, have adopted a friendly stance toward blockchain technology, while others, such as China and India, have banned it altogether or placed heavy restrictions on its use.

The lack of global regulations and the differing opinions of governments and regulators on blockchain technology have created an uncertain environment for investors and businesses. However, many experts believe that regulations are necessary to ensure the legitimacy and stability of the blockchain market.

In addition to regulations, the future of blockchain technology is likely to be shaped by technological advancements and innovation. Many experts predict that the blockchain will continue to evolve and adapt to changing market conditions, leading to the development of new applications and use cases beyond the cryptocurrency market.

One area of innovation that is already being explored is the use of smart contracts on the blockchain. Smart contracts are self-executing contracts with the terms of the agreement directly written into code. They can be used to automate complex processes, such as financial transactions or supply chain management, and have the potential to revolutionize many industries.

Another area of innovation is the development of new blockchain platforms and protocols. Ethereum, for example, is a blockchain platform that allows for the creation of decentralized applications, or dApps. These dApps can be used for a wide range of purposes, from decentralized finance to social media.

In conclusion, blockchain technology is a decentralized, digital ledger that provides a secure and tamper-proof system for storing and transferring digital assets. It is at the heart of the cryptocurrency revolution and has the potential to revolutionize many other industries beyond finance.

Despite the challenges and controversies associated with blockchain technology, such as scalability and regulation, its benefits are clear. It provides a secure and transparent system for transactions, and its potential for innovation and disruption is enormous.

As the blockchain continues to evolve and adapt, it is important for investors and businesses to stay informed and educated about this new technology. With its unique features and potential benefits, blockchain technology has the power to transform the way we conduct business and interact with each other in the digital age.

The blockchain is still a relatively new technology, and its full potential has yet to be realized. However, the widespread adoption of cryptocurrencies and the growing interest in blockchain technology suggest that it is here to stay. As more businesses and industries explore the potential of the blockchain, it is likely that we will see new and innovative use cases emerge.

One area where the blockchain has already made significant strides is in the realm of financial services. The rise of decentralized finance, or DeFi, has demonstrated the potential of the blockchain to create new financial products and services that are more accessible and transparent than traditional finance.

DeFi platforms are built on the blockchain and offer a wide range of financial services, including lending, borrowing, and trading. These platforms are decentralized, meaning that they are not controlled by any central authority, and allow users to transact with each other directly, without the need for intermediaries.

The growth of DeFi has been fueled by the popularity of cryptocurrencies, which have made it easier than ever to transact across borders and to access financial services that were previously unavailable to many people. With the use of blockchain technology, DeFi platforms have been able to create new financial products that are more accessible, transparent, and secure than traditional finance.

In addition to DeFi, the blockchain has the potential to disrupt many other industries, including supply chain management, healthcare, and voting. The immutability and transparency of the blockchain make it an ideal platform for creating secure and tamper-proof systems that can be used to store and share sensitive information.

For example, in the healthcare industry, the blockchain can be used to create secure systems for storing and sharing medical records. This would allow patients to have more control over their health data and make it easier for healthcare providers to access and share information securely.

Similarly, the blockchain can be used to create secure voting systems that are resistant to fraud and manipulation. This would increase the trust and transparency of the electoral process, and could help to restore faith in democracy.

Conclusion

The rise of blockchain technology has the potential to revolutionize many aspects of our lives, from finance to healthcare to voting. Its decentralized and transparent nature makes it an ideal platform for creating secure and tamper-proof systems that can be used to store and share information.

As the blockchain continues to evolve and adapt, it is important for investors and businesses to stay informed and educated about this new technology. With its unique features and potential benefits, blockchain technology has the power to transform the way we conduct business and interact with each other in the digital age.

Chapter 3

"Bitcoin: The Original Cryptocurrency and Its Impact on the World of Finance"

Bitcoin is the world's first cryptocurrency, and it has had a profound impact on the world of finance since its creation in 2009. It has disrupted traditional banking systems, introduced new forms of digital currency, and sparked a global movement toward decentralization and financial independence. In this essay, we will explore the origins of Bitcoin, how it works, and its impact on the world of finance.

Origins of Bitcoin

Bitcoin was created in 2009 by an unknown person or group using the pseudonym Satoshi Nakamoto. The goal was to create a decentralized, digital currency that would allow for secure and private transactions without the need for intermediaries, such as banks or governments.

The creation of Bitcoin was made possible by a new technology called blockchain. The blockchain is a decentralized digital ledger that records every transaction made using the currency. Each block in the blockchain contains a unique cryptographic code, and once a block is added to the chain, it cannot be altered. This makes the blockchain a tamper-proof and secure system for storing and transferring digital assets.

How Bitcoin Works

Bitcoin works by using a decentralized network of computers to validate and verify transactions. Each transaction is recorded on the blockchain, and once it is added to the chain, it cannot be altered or deleted.

To participate in the Bitcoin network, users need to have a digital wallet that contains their Bitcoin holdings. Transactions are initiated by sending Bitcoin from one wallet to another, and each transaction is validated by the network of computers on the blockchain.

When a new transaction is initiated, it is broadcast to the network of computers on the blockchain. Each computer on the network verifies the transaction to ensure that it is valid and has not already been recorded on the blockchain. Once the transaction is verified, it is added to a new block, which is then broadcast to the network for validation.

Once a block is validated by the network, it is added to the blockchain, and the transaction is considered complete. The transaction is now visible to anyone on the network, and the ledger is updated to reflect the new transaction.

Impact on the World of Finance

Bitcoin has had a significant impact on the world of finance since its creation in 2009. It has disrupted traditional banking systems, introduced new forms of digital currency, and sparked a global movement toward decentralization and financial independence.

One of the main benefits of Bitcoin is its decentralized nature. Unlike traditional currencies that are controlled by governments and banks, Bitcoin is not subject to central control. This means that it is not affected by inflation, interest rates, or other economic factors that can impact traditional currencies. Instead, the value of Bitcoin is determined by the market, based on supply and demand.

Another benefit of Bitcoin is its security. The blockchain technology that underpins Bitcoin is highly secure and difficult to hack. This makes Bitcoin an attractive option for those who want to keep their financial transactions private and secure.

Bitcoin has also introduced new forms of digital currency, such as altcoins, which are similar to Bitcoin but have their own unique features and characteristics. Altcoins have gained popularity in recent years and have created new opportunities for investors and businesses.

Challenges and Controversies

Despite the many benefits of Bitcoin, there are also challenges and controversies associated with it. One of the biggest challenges facing Bitcoin is its volatility. The value of Bitcoin can fluctuate rapidly, making it a risky investment for those who are not familiar with the market.

Bitcoin has also been associated with a number of controversies, including its use in illegal activities such as money laundering and terrorism financing. Additionally, some critics argue that Bitcoin is not backed by any tangible assets and that its value is based solely on speculation, making it a highly speculative and volatile investment.

Furthermore, the lack of regulation and oversight in the Bitcoin market has led to concerns from governments and regulators around the world. Some countries, such as China and India, have banned Bitcoin altogether, while others, such as the United States, have implemented regulations to mitigate the risks associated with cryptocurrencies.

Regulations and Future of Bitcoin

As the use and popularity of Bitcoin continue to grow, governments and regulators around the world are grappling with how to regulate this new form of currency. While some countries have banned Bitcoin outright, others have adopted a more friendly stance toward cryptocurrencies and blockchain technology.

In the United States, for example, the Securities and Exchange Commission (SEC) has taken steps to regulate the sale and trading of cryptocurrencies. The SEC has also created a framework for the classification of cryptocurrencies, which has helped to provide clarity and guidance for investors and businesses.

Despite the challenges and controversies associated with Bitcoin, its future is bright. The widespread adoption of Bitcoin and other cryptocurrencies has sparked a global movement toward decentralization and financial independence. As more businesses and industries explore the potential of blockchain technology, it is likely that we will see new and innovative use cases emerge.

Conclusion

In conclusion, Bitcoin is the world's first cryptocurrency and has had a profound impact on the world of finance since its creation in 2009. It has disrupted traditional banking systems, introduced new forms of digital currency, and sparked a global movement toward decentralization and financial independence.

Despite the challenges and controversies associated with Bitcoin, its benefits are clear. It provides a decentralized and secure system for storing and transferring digital assets, and its potential for innovation and disruption is enormous.

As the Bitcoin market continues to evolve and adapt, it is important for investors and businesses to stay informed and educated about this new form of currency. With its unique features and potential benefits, Bitcoin has the power to transform the way we think about money and financial transactions.

Chapter 4

"The Pros and Cons of Investing in Cryptocurrencies: A Comprehensive Analysis"

Cryptocurrencies have become increasingly popular in recent years, with Bitcoin leading the way as the world's first and most well-known cryptocurrency. While cryptocurrencies offer many potential benefits, such as decentralization and security, they also come with significant risks and challenges. In this essay, we will explore the pros and cons of investing in cryptocurrencies, providing a comprehensive analysis of the benefits and drawbacks of this new asset class.

Pros:

Decentralization: Cryptocurrencies are decentralized, which means they are not controlled by any central authority. This provides users with greater control over their financial transactions and reduces the risk of interference or manipulation. Additionally, decentralization provides a higher degree of privacy and anonymity, which can be important for some users.

Security: Cryptocurrencies are secured by advanced cryptographic algorithms, which make them virtually impossible to hack or counterfeit. This provides users with a high level of security and protection from fraud. Additionally, the use of blockchain technology ensures that transactions are transparent and tamper-proof, providing further security and accountability.

Transparency: The blockchain technology that underpins cryptocurrencies provides transparency and accountability. The public ledger records every transaction made using the currency, and once a transaction is added to the blockchain, it cannot be altered or deleted. This makes it easy for users to track their transactions and ensures that the system is fair and honest.

Potential for high returns: Cryptocurrencies have the potential to provide investors with high returns on their investments, particularly in the short term. The value of cryptocurrencies can fluctuate rapidly, and savvy investors can take advantage of these fluctuations to make a profit. Additionally, the growing popularity of cryptocurrencies and the increasing number of businesses that accept them as payment suggests that they may continue to increase in value over time.

Access to new markets: Cryptocurrencies provide investors with access to new and emerging markets that may not be available through traditional investment channels. This can provide diversification and new opportunities for growth. Additionally, cryptocurrencies can be traded 24/7, providing investors with greater flexibility and control over their investments.

Cons:

Volatility: Cryptocurrencies are highly volatile, with their value fluctuating rapidly and unpredictably. This can make them a risky investment, particularly for inexperienced investors. The value of cryptocurrencies can be influenced by a wide range of factors, such as market sentiment, news events, and regulatory changes.

Lack of regulation: Cryptocurrencies are largely unregulated, which can make them susceptible to fraud, manipulation, and other illegal activities. Additionally, the lack of regulation can make it difficult for investors to assess the risks and make informed investment decisions.

Lack of acceptance: Cryptocurrencies are not widely accepted as a form of payment, and it can be difficult to find merchants and businesses that accept them. This can limit their usefulness and adoption, particularly for investors who are looking for a practical and convenient means of payment.

Cybersecurity risks: Cryptocurrencies are vulnerable to cybersecurity risks, such as hacking and theft. This can result in the loss of investments and other assets. Additionally, the lack of regulation and oversight in the cryptocurrency market can make it difficult for investors to recover their losses in the event of a cybersecurity breach.

Limited understanding: Cryptocurrencies are a new and complex asset class, and many investors may not fully understand how they work or the risks and challenges associated with them. This can lead to uninformed investment decisions and increase the risk of losses.

Conclusion

Investing in cryptocurrencies can provide significant benefits, such as decentralization, security, and potential high returns. However, it also comes with significant risks and challenges, such as volatility, lack of regulation, cybersecurity risks, and limited understanding.

Investors who are considering investing in cryptocurrencies should carefully weigh the pros and cons and seek out expert advice and guidance before making any investment decisions. While cryptocurrencies offer many potential benefits, they are not suitable for all investors, and careful consideration and due diligence are required to make informed investment decisions.

As the cryptocurrency market continues to evolve and adapt, it is important for investors to stay informed and educated about this new

Chapter 5

Altcoins: Exploring the Diversity of Cryptocurrencies Beyond Bitcoin

When most people think of cryptocurrencies, Bitcoin is usually the first currency that comes to mind. However, Bitcoin is just one of many cryptocurrencies that have emerged in recent years. These alternative cryptocurrencies, or altcoins, offer a diverse range of features and use cases beyond what Bitcoin can offer. In this essay, we will explore the world of altcoins, the benefits they offer, and the challenges they face.

What are Altcoins?

Altcoins are alternative cryptocurrencies to Bitcoin that were created in the wake of Bitcoin's success. They use similar blockchain technology to Bitcoin but have their own unique features and uses. Altcoins have evolved from simple clones of Bitcoin to more sophisticated and innovative cryptocurrencies that offer features such as faster transaction times, enhanced privacy, and smart contracts.

Some of the most popular altcoins include Ethereum, Litecoin, Ripple, and Bitcoin Cash. Each altcoin has its own unique set of features and use cases that distinguish it from Bitcoin and other cryptocurrencies.

Benefits of Altcoins

Altcoins offer a number of benefits over Bitcoin and other traditional forms of currency. Some of the most notable benefits of altcoins include:

Diversification: Altcoins provide investors with a diverse range of investment opportunities beyond Bitcoin. This can help to reduce the risk of investment and provide more options for investors to choose from.

Innovation: Altcoins are at the forefront of cryptocurrency innovation, with many cryptocurrencies offering unique features such as smart contracts, faster transaction times, and enhanced privacy. This innovation helps to push the boundaries of what is possible with cryptocurrencies and can lead to new use cases and applications.

Community: Many altcoins have vibrant and engaged communities that are passionate about their chosen currency. These communities can provide support, resources, and education to users, making it easier to learn about and use altcoins.

Use Cases: Altcoins can be used for a variety of purposes beyond traditional payments, such as for smart contracts, decentralized applications, and other innovative use cases. This versatility makes altcoins more useful and adaptable to a wide range of industries and applications.

Challenges of Altcoins

While altcoins offer many benefits, they also face a number of challenges and risks. Some of the most notable challenges of altcoins include:

Lack of Adoption: Altcoins can struggle with adoption, as they are not as well-known or widely used as Bitcoin. This can limit their usefulness and adoption, particularly for investors who are looking for a practical and convenient means of payment.

Volatility: Altcoins are highly volatile, with their value fluctuating rapidly and unpredictably. This can make them a risky investment, particularly for inexperienced investors. The value of altcoins can be influenced by a wide range of factors, such as market sentiment, news events, and regulatory changes.

Security: Altcoins can be vulnerable to security risks such as hacking, phishing, and other forms of cyber-attacks. Additionally, altcoin exchanges can be vulnerable to insider threats and fraud, which can result in the loss of investments and other assets.

Regulation: Altcoins are largely unregulated, which can make them susceptible to fraud, manipulation, and other illegal activities. Additionally, the lack of regulation can make it difficult for investors to assess the risks and make informed investment decisions.

Conclusion

Altcoins offer a diverse range of features and use cases beyond what Bitcoin can offer. They provide investors with a diverse range of investment opportunities, push the boundaries of what is possible with cryptocurrencies, and can be used for a variety of innovative applications.

However, altcoins also face a number of challenges and risks, including lack of adoption, volatility, security, and regulation. To address these challenges, altcoins will need to continue to innovate, build strong communities, and work with regulators to establish a clear and effective regulatory framework.

One area where altcoins are already making significant progress is in the development of decentralized finance (DeFi) applications. DeFi is a growing sector of the cryptocurrency industry that seeks to decentralize traditional financial services such as lending, borrowing, and trading. Altcoins such as Ethereum and Binance Coin are leading the way in the development of DeFi applications, which could have significant implications for the future of finance.

Overall, altcoins offer a diverse and dynamic landscape of cryptocurrencies that are pushing the boundaries of what is possible with digital assets. While they face challenges and risks, they also offer a range of benefits and opportunities for investors, developers, and users. As the cryptocurrency industry continues to evolve, altcoins will play an important role in shaping the future of finance and technology.

Chapter 6

"Crypto Regulations Around the World: How Governments are Responding to the Rise of Cryptocurrencies"

The rise of cryptocurrencies has presented governments and regulators around the world with a unique set of challenges. On one hand, cryptocurrencies offer a number of potential benefits, such as decentralization and security. On the other hand, they also come with significant risks, such as volatility and susceptibility to illegal activities such as money laundering and terrorism financing. In this essay, we will explore how governments around the world are responding to the rise of cryptocurrencies, and the different approaches they are taking to regulate this new and complex asset class.

North America

The United States and Canada have taken different approaches to regulating cryptocurrencies. In the United States, the Securities and Exchange Commission (SEC) has taken steps to regulate the sale and trading of cryptocurrencies. The SEC has also created a framework for the classification of cryptocurrencies, which has helped to provide clarity and guidance for investors and businesses.

In Canada, the government has taken a more hands-off approach to regulating cryptocurrencies. While the Canadian government has not yet implemented any formal regulations for cryptocurrencies, the Financial Consumer Agency of Canada (FCAC) has issued warnings to consumers about the risks associated with investing in cryptocurrencies.

Europe

Europe has been at the forefront of cryptocurrency regulation, with several countries implementing formal regulations for cryptocurrencies. In 2018, the European Union introduced new regulations for cryptocurrencies, requiring exchanges to register with national authorities and to conduct due diligence on their customers.

Several European countries have also introduced their own regulations for cryptocurrencies. In Germany, for example, cryptocurrencies are considered a financial instrument and are subject to the same regulations as traditional financial instruments. In France, cryptocurrency exchanges are required to register with the government and comply with anti-money laundering regulations.

Asia

Asia has a complex regulatory landscape for cryptocurrencies, with some countries taking a friendly approach to cryptocurrencies and others outright banning them. Japan, for example, has adopted a friendly stance toward cryptocurrencies, with the government recognizing Bitcoin as a legal form of payment. Japan has also implemented a licensing system for cryptocurrency exchanges and has strict regulations in place to protect consumers.

China, on the other hand, has banned the use of cryptocurrencies altogether. In 2017, the Chinese government banned initial coin offerings (ICOs), citing concerns about fraud and illegal fundraising. Additionally, China has implemented strict regulations on cryptocurrency exchanges and mining operations.

Other countries in Asia, such as South Korea and Singapore, have taken a more nuanced approach to regulating cryptocurrencies. South Korea, for example, has introduced regulations to protect consumers and prevent money laundering, while also recognizing the potential benefits of cryptocurrencies. Singapore has implemented regulations to prevent money laundering and terrorism financing, while also supporting innovation and growth in the cryptocurrency industry.

Middle East

The Middle East has been relatively slow to adopt cryptocurrencies, with many countries taking a cautious approach to regulation. In 2019, the United Arab Emirates (UAE) introduced new regulations for cryptocurrency exchanges, requiring them to obtain a license from the government and comply with anti-money laundering regulations.

Other countries in the Middle East, such as Saudi Arabia and Kuwait, have not yet implemented any formal regulations for cryptocurrencies. However, the governments of these countries have issued warnings to consumers about the risks associated with investing in cryptocurrencies.

Africa

Africa is a relatively new and emerging market for cryptocurrencies, with several countries taking different approaches to regulating this new asset class. South Africa, for example, has introduced regulations to prevent money laundering and terrorism financing, while also supporting innovation and growth in the cryptocurrency industry.

Other countries in Africa, such as Nigeria and Ghana, have not yet implemented any formal regulations for cryptocurrencies. However, the governments of these countries have expressed interest in exploring the potential of blockchain technology and cryptocurrencies, and have launched initiatives to promote innovation and growth in this space.

Australia

Australia has taken a cautious approach to regulating cryptocurrencies, with the government introducing a number of measures to prevent money laundering and terrorism financing. In 2018, the Australian government introduced new regulations for cryptocurrency exchanges, requiring them to register with the Australian Transaction Reports and Analysis Centre (AUSTRAC) and comply with anti-money laundering and counter-terrorism financing regulations.

Additionally, the Australian Securities and Investments Commission (ASIC) has issued guidance on initial coin offerings (ICOs) and has warned consumers about the risks associated with investing in cryptocurrencies. The Australian government has also launched initiatives to promote innovation and growth in the blockchain and cryptocurrency industry, such as the Digital Transformation Agency's Blockchain Roadmap and the Australian Securities Exchange's (ASX) blockchain-based trading platform.

Conclusion

The regulatory landscape for cryptocurrencies is complex and rapidly evolving, with different countries taking different approaches to regulating this new and complex asset class. While some countries have embraced cryptocurrencies and blockchain technology, others have taken a more cautious or outright hostile approach.

As the use and popularity of cryptocurrencies continue to grow, it is likely that we will see more countries implementing formal regulations for cryptocurrencies. These regulations will need to strike a balance between protecting consumers and preventing illegal activities, while also promoting innovation and growth in the cryptocurrency industry.

Investors and businesses operating in the cryptocurrency industry will need to stay informed and educated about the different regulatory environments around the world, and take steps to ensure compliance with local regulations. Additionally, they will need to be prepared to adapt to changing regulations and market conditions, and to embrace innovation and new opportunities in this rapidly evolving industry.

Chapter 7

"The Future of Crypto: Predictions and Trends for the Next Decade"

The cryptocurrency industry has come a long way since the launch of Bitcoin in 2009. Today, there are thousands of cryptocurrencies, with a total market capitalization of over $1.1 trillion. As the industry continues to evolve and mature, there are a number of predictions and trends that are likely to shape the future of crypto in the coming decade. In this essay, we will explore some of these predictions and trends, and what they could mean for the future of crypto.

Prediction 1: Increased Institutional Adoption

One of the most significant trends in the crypto industry over the past few years has been the increasing adoption of cryptocurrencies by institutional investors. This trend is likely to continue in the coming decade, as more traditional financial institutions begin to recognize the value and potential of cryptocurrencies.

Already, major financial institutions such as JPMorgan and Goldman Sachs have launched cryptocurrency trading desks, and major investment firms such as BlackRock have begun investing in cryptocurrencies. As more institutions enter the market, they are likely to bring greater liquidity and stability to the crypto market, making it more attractive to investors and businesses.

Prediction 2: Greater Integration with Traditional Finance

Another trend that is likely to shape the future of crypto is greater integration with traditional finance. As the cryptocurrency industry continues to mature, it is likely that we will see greater collaboration and partnerships between traditional financial institutions and cryptocurrency firms.

Already, we are seeing this trend emerge in the form of cryptocurrencies being used as collateral for loans, and in the development of decentralized finance (DeFi) applications. In the coming years, we are likely to see more integration between the cryptocurrency industry and traditional finance, with cryptocurrencies being used for a wider range of financial services and applications.

Prediction 3: Increased Government Regulation

As cryptocurrencies continue to grow in popularity and market capitalization, it is likely that we will see increased government regulation of the industry. While some countries have already implemented formal regulations for cryptocurrencies, such as the European Union's MiCA proposal, many countries have yet to establish clear and effective regulatory frameworks for cryptocurrencies.

However, as the industry continues to mature, it is likely that governments will begin to take a more active role in regulating cryptocurrencies. This could include the introduction of licensing requirements for cryptocurrency exchanges and stricter regulations around the use of cryptocurrencies for illegal activities such as money laundering and terrorism financing.

Prediction 4: Greater Focus on Privacy

One of the key features of cryptocurrencies is their ability to offer greater privacy and anonymity than traditional forms of payment. However, this feature has also made cryptocurrencies attractive to criminals and other illegal activities.

In the coming decade, it is likely that we will see a greater focus on privacy in the cryptocurrency industry. This could include the development of new privacy-focused cryptocurrencies, as well as the introduction of privacy features for existing cryptocurrencies.

Prediction 5: Continued Innovation in Blockchain Technology

Finally, it is likely that we will see continued innovation in blockchain technology in the coming decade. While blockchain technology was originally developed to power cryptocurrencies, its potential applications go far beyond this.

In the coming years, we are likely to see greater use of blockchain technology in a wide range of industries, from healthcare to supply chain management. This could include the development of blockchain-based solutions for data management, identity verification, and other innovative applications.

Conclusion

The future of crypto is likely to be shaped by a range of predictions and trends, from increased institutional adoption and greater integration with traditional finance to increased government regulation and a greater focus on privacy. Additionally, continued innovation in blockchain technology is likely to open up new applications and opportunities for the industry.

As the cryptocurrency industry continues to evolve and mature, it is likely that we will see a growing range of use cases and applications for cryptocurrencies and blockchain technology. From decentralized finance to digital identity and asset management, the possibilities are virtually endless.

While there are certainly challenges and risks associated with the cryptocurrency industry, the potential benefits and opportunities are too significant to ignore. Whether you are an investor, developer, or simply a curious observer, the future of crypto is sure to be an exciting and dynamic space to watch.

As we move forward into the next decade, it will be important for stakeholders across the industry to work together to address the challenges and realize the potential of cryptocurrencies and blockchain technology. Whether through collaboration with governments and regulators, the development of new and innovative applications, or the promotion of greater education and awareness, there is much that can be done to help shape the future of crypto in a positive and impactful way.

Overall, the future of crypto is one that is full of potential and possibility. While there are certainly challenges and risks associated with this new and emerging asset class, there is also much to be excited about. As the industry continues to evolve and mature, it is up to all of us to work together to ensure that cryptocurrencies and blockchain technology are used in ways that benefit society as a whole. Whether you are a long-time supporter of the industry or simply curious about what the future holds, there has never been a better time to get involved in the exciting world of crypto.

Chapter 8

"DeFi: The Emergence of Decentralized Finance and its Relationship to Cryptocurrencies"

Decentralized finance, or DeFi, is a rapidly growing sector of the cryptocurrency industry that seeks to decentralize traditional financial services such as lending, borrowing, and trading. DeFi applications are built on blockchain technology, allowing for greater transparency, security, and accessibility than traditional financial services. In this essay, we will explore the emergence of DeFi, its relationship to cryptocurrencies, and its potential impact on the future of finance.

What is DeFi?

DeFi refers to a set of decentralized financial applications that operate on blockchain technology. These applications are designed to provide users with greater control over their finances, while also offering greater transparency and security than traditional financial services. Some of the most popular DeFi applications include decentralized exchanges (DEXs), lending and borrowing platforms, and stablecoins.

One of the key features of DeFi applications is that they are open and accessible to anyone with an internet connection, regardless of their geographic location or financial status. This has the potential to democratize financial services and provide greater access to financial opportunities for those who have traditionally been excluded from the traditional financial system.

DeFi and Cryptocurrencies

While DeFi applications are built on blockchain technology, they are not necessarily the same as cryptocurrencies. However, cryptocurrencies have played a key role in the development of DeFi, providing the means of exchange and store of value necessary for these applications to function.

Many DeFi applications rely on stablecoins, which are cryptocurrencies that are pegged to a stable asset such as the US dollar. Stablecoins provide the stability and predictability necessary for lending and borrowing platforms to operate effectively, as they reduce the volatility and risk associated with traditional cryptocurrencies such as Bitcoin and Ethereum.

Additionally, cryptocurrencies are often used as collateral for loans on DeFi platforms, allowing users to access liquidity without the need for traditional financial intermediaries. This can help to reduce costs and increase accessibility for borrowers, while also providing lenders with a new and potentially lucrative source of income.

Benefits of DeFi

DeFi offers a number of benefits over traditional financial services, including:

Decentralization: DeFi applications are decentralized, meaning that they operate on a distributed network of computers rather than a central server or institution. This reduces the risk of censorship, fraud, and other forms of malfeasance.

Transparency: DeFi applications are transparent, with all transactions and interactions recorded on a public blockchain. This provides greater transparency and accountability than traditional financial services.

Accessibility: DeFi applications are accessible to anyone with an internet connection, regardless of their geographic location or financial status. This has the potential to democratize financial services and provide greater access to financial opportunities for those who have traditionally been excluded from the traditional financial system.

Efficiency: DeFi applications are often more efficient than traditional financial services, as they rely on automated processes and smart contracts rather than manual processes and intermediaries. This can help to reduce costs and increase speed and accuracy.

Challenges of DeFi

While DeFi offers a number of benefits, it also faces a number of challenges and risks. Some of the most notable challenges of DeFi include:

Security: DeFi applications can be vulnerable to security risks such as hacking, phishing, and other forms of cyber attacks. Additionally, DeFi protocols can be vulnerable to code exploits and other forms of technical vulnerabilities.

Regulation: DeFi is largely unregulated, which can make it susceptible to fraud, manipulation, and other illegal activities. Additionally, the lack of regulation can make it difficult for users to assess the risks and make informed investment decisions.

Volatility: While stablecoins can provide stability and predictability for DeFi applications, other cryptocurrencies can be highly volatile. This can make it difficult for users to accurately assess the risks and returns of DeFi investments, and can also result in significant losses for investors.

User Experience: DeFi applications can be complex and difficult to use, especially for those who are not familiar with blockchain technology. This can make it challenging for DeFi to reach a wider audience and achieve mainstream adoption.

Despite these challenges, the potential benefits of DeFi are significant, and the industry is expected to continue to grow and evolve in the coming years.

Impact of DeFi on the Future of Finance

The emergence of DeFi has the potential to revolutionize the way that financial services are provided, making them more accessible, transparent, and efficient. This could have significant implications for the future of finance, as traditional financial institutions and services are disrupted and decentralized.

Some of the potential impacts of DeFi on the future of finance include:

Democratization of Financial Services: DeFi has the potential to provide greater access to financial services for individuals and businesses that have traditionally been excluded from the traditional financial system. This could help to reduce inequality and increase economic opportunity for all.

Disintermediation: DeFi could help to disintermediate traditional financial intermediaries such as banks, reducing costs and increasing efficiency for users.

Innovation: DeFi is a highly innovative industry, with new applications and use cases being developed all the time. This could lead to the development of new and innovative financial services that are not currently available through traditional financial institutions.

Regulatory Challenges: As DeFi continues to grow and mature, it is likely that it will face increasing regulatory scrutiny. This could present challenges for the industry, as it navigates complex and often conflicting regulatory requirements.

Conclusion

Decentralized finance, or DeFi, is a rapidly growing sector of the cryptocurrency industry that seeks to decentralize traditional financial services such as lending, borrowing, and trading. DeFi applications are built on blockchain technology, offering greater transparency, security, and accessibility than traditional financial services. While DeFi faces a number of challenges and risks, the potential benefits are significant, and the industry is expected to continue to grow and evolve in the coming years.

As DeFi continues to mature, it has the potential to revolutionize the way that financial services are provided, making them more accessible, transparent, and efficient. This could have significant implications for the future of finance, as traditional financial institutions and services are disrupted and decentralized.

Overall, DeFi is an exciting and dynamic industry that is pushing the boundaries of what is possible with digital assets and blockchain technology. Whether you are an investor, developer, or simply a curious observer, there has never been a better time to get involved in the exciting world of DeFi.

Chapter 9

"Crypto Security: Best Practices for Keeping Your Digital Assets Safe"

Cryptocurrencies offer a new and exciting way to store and transfer value, but they also come with unique security challenges. Unlike traditional forms of currency, cryptocurrencies are digital assets that are stored on a blockchain, which is a decentralized and distributed ledger. This means that the security of your crypto assets is largely dependent on your own actions and best practices. In this essay, we will explore some of the best practices for keeping your digital assets safe and secure.

Use a Hardware Wallet

One of the best ways to keep your digital assets safe is to use a hardware wallet. A hardware wallet is a physical device that stores your private keys and allows you to securely sign transactions. By keeping your private keys offline, hardware wallets provide a high level of security against online threats such as hacking and phishing.

There are several popular hardware wallets available on the market, including the Ledger Nano S and the Trezor. When setting up your hardware wallet, be sure to follow the manufacturer's instructions carefully and keep your recovery phrase in a secure location.

Use Strong Passwords and Two-Factor Authentication

Another important best practice for crypto security is to use strong passwords and two-factor authentication. When creating a password for your crypto wallet or exchange account, be sure to use a strong and unique password that is not used for any other accounts. A strong password should be at least 12 characters long and include a combination of letters, numbers, and special characters.

In addition to using strong passwords, it is also important to enable two-factor authentication (2FA) wherever possible. 2FA adds an additional layer of security by requiring a second form of authentication, such as a code sent to your phone or a biometric scan.

Keep Your Software Up to Date

Another important best practice for crypto security is to keep your software up to date. This includes your operating system, web browser, and any crypto-related software such as wallets and exchanges. Software updates often include security patches and bug fixes that can help to protect your digital assets against online threats.

Be sure to regularly check for updates and install them as soon as they become available. Additionally, be wary of downloading software or apps from untrusted sources, as these may contain malware or other security vulnerabilities.

Avoid Public Wi-Fi Networks

Using public Wi-Fi networks can be convenient, but it can also put your digital assets at risk. Public Wi-Fi networks are often unsecured, meaning that anyone on the same network can potentially access your data and steal your private keys.

When accessing your crypto wallet or exchange account, be sure to use a secure and trusted network. If you must use a public Wi-Fi network, consider using a virtual private network (VPN) to encrypt your data and protect your privacy.

Be Wary of Phishing Scams

Phishing scams are a common tactic used by cybercriminals to steal personal information and access to your digital assets. Phishing scams typically involve sending fake emails or messages that appear to be from a trusted source, such as your crypto wallet or exchange.

To protect yourself from phishing scams, be wary of any unsolicited emails or messages that ask for your personal information or login credentials. Additionally, be sure to double-check the URL of any website you are visiting, as phishing websites often use URLs that are similar to legitimate websites.

Diversify Your Crypto Holdings

Finally, it is important to diversify your crypto holdings as a best practice for crypto security. By spreading your investments across multiple cryptocurrencies and exchanges, you can help to mitigate the risks of any single point of failure.

Additionally, be sure to regularly review your crypto holdings and adjust your portfolio as needed. This can help to ensure that your investments are aligned with your risk tolerance and financial goals.

Conclusion

In conclusion, crypto security is a critical component of any crypto investment strategy. By following these best practices, you can help to protect your digital assets against online threats and ensure that your investments are safe and secure.

Remember to use a hardware wallet, use strong passwords and two-factor authentication, keep your software up to date, avoid public Wi-Fi networks, be wary of phishing scams, diversify your crypto holdings, and regularly review your portfolio. By incorporating these best practices into your crypto investment strategy, you can help to protect your digital assets and maximize your returns.

Overall, the crypto industry is still in its early stages, and security is an ongoing concern. As the industry continues to evolve and mature, it will be important for investors and businesses to stay vigilant and take proactive steps to protect their digital assets. With the right tools and strategies, however, it is possible to invest in cryptocurrencies safely and securely, and to reap the potential benefits of this exciting and dynamic asset class.

Chapter 10

"The Psychology of Crypto: Understanding the Emotions and Behaviors of Crypto Investors".

Investing in cryptocurrencies can be a highly emotional experience, driven by fear, greed, and a desire for financial freedom. Crypto investors are often characterized by their willingness to take risks, their belief in the potential of digital assets, and their passion for technology and innovation. In this essay, we will explore the psychology of crypto and the emotions and behaviors that drive crypto investors.

Fear and Greed

Fear and greed are two of the most powerful emotions that drive crypto investors. Fear can manifest as a fear of missing out (FOMO), a fear of losing money, or a fear of being left behind in a rapidly evolving industry. Greed, on the other hand, can manifest as a desire for quick profits, a desire for financial freedom, or a desire to be part of a new and exciting movement.

These emotions can be heightened by the extreme volatility of the crypto market, which can cause prices to fluctuate rapidly and dramatically. When prices are rising, investors may be driven by greed, as they try to ride the wave of a bull market. When prices are falling, investors may be driven by fear, as they try to protect their investments and avoid losses.

Confirmation Bias

Confirmation bias is a cognitive bias that can impact the decisions of crypto investors. This bias occurs when individuals seek out information that confirms their pre-existing beliefs or opinions, while ignoring information that contradicts them.

In the context of crypto investing, confirmation bias can manifest as a tendency to seek out positive news and information about a particular cryptocurrency, while ignoring negative news and information. This can lead to overconfidence and a false sense of security, as investors may fail to consider the risks and uncertainties associated with their investments.

Herd Mentality

Herd mentality is another psychological factor that can impact the decisions of crypto investors. This refers to the tendency of individuals to follow the crowd and conform to the opinions and behaviors of others, rather than making independent decisions.

In the context of crypto investing, herd mentality can manifest as a tendency to follow the investment decisions of others, rather than conducting independent research and analysis. This can lead to a situation where investors all pile into a particular cryptocurrency, driving up prices in a speculative bubble, only to see prices crash when the bubble bursts.

Cognitive Biases

In addition to confirmation bias, there are a number of other cognitive biases that can impact the decisions of crypto investors. These include:

Anchoring bias: the tendency to rely too heavily on the first piece of information encountered when making a decision.

Overconfidence bias: the tendency to overestimate one's own abilities and knowledge.

Endowment bias: the tendency to value something more highly simply because you own it.

Sunk cost fallacy: the tendency to continue investing in something simply because you have already invested a lot of money in it, even if it no longer makes logical sense to do so.

Future Focus

The future focus of crypto investors is driven by the potential for blockchain technology to transform industries and disrupt traditional systems. Crypto investors believe that digital assets and decentralized networks have the potential to democratize finance, increase transparency and accountability, and unlock new opportunities for innovation.

This future focus is reflected in the many ambitious projects and initiatives that are being developed in the crypto industry. From decentralized finance (DeFi) to non-fungible tokens

(NFTs) and beyond, crypto investors are constantly pushing the boundaries of what is possible with digital assets and blockchain technology.

However, it is important to remember that the future of crypto is not without risks and challenges. As with any emerging technology, there are many uncertainties and potential pitfalls that must be considered. These include regulatory challenges, security risks, and scalability issues, among others.

To succeed in the world of crypto investing, it is important to balance a future focus with a realistic assessment of the risks and challenges that lie ahead. This means conducting thorough research, staying up to date on industry developments, and carefully evaluating the potential risks and rewards of any investment opportunity.

In addition, it is important to remember that the future of crypto is not predetermined. While digital assets and blockchain technology have the potential to transform industries and disrupt traditional systems, their success is ultimately dependent on a range of factors, including regulatory frameworks, adoption rates, and market demand.

Conclusion

In conclusion, the future focus of crypto investors is driven by a belief in the potential of digital assets and blockchain technology to transform industries and disrupt traditional systems. While this future focus is a positive force for innovation and investment, it is important to balance it with a realistic assessment of the risks and challenges that lie ahead.

To navigate the world of crypto investing successfully, it is important to stay informed, conduct thorough research, and carefully evaluate the potential risks and rewards of any investment opportunity. By doing so, investors can position themselves for success in the dynamic and rapidly evolving world of cryptocurrencies.

One area of innovation that has emerged in recent years is decentralized finance (DeFi). DeFi refers to a range of financial applications that use blockchain technology to create new financial instruments and services that are more accessible and transparent than traditional financial systems.

DeFi projects often rely on cryptocurrencies to operate, and have been growing rapidly in recent years. One example of a DeFi project is decentralized exchanges (DEXs), which allow users to trade cryptocurrencies without the need for a centralized exchange. This can provide greater transparency and security, as well as lower fees compared to traditional exchanges.

Another area of innovation is non-fungible tokens (NFTs), which are unique digital assets that are verified on a blockchain. NFTs can be used to represent a variety of assets, such as art, music, and collectibles. They have become increasingly popular in recent years, with some NFTs selling for millions of dollars.

As the crypto industry continues to mature, it is likely that we will see new and innovative use cases emerge, as well as further developments in blockchain technology. However, it will also be important for regulators and stakeholders to address the challenges and risks associated with cryptocurrencies, such as volatility, security, and regulatory uncertainty.

Overall, cryptocurrencies are a promising and exciting development in the world of finance and technology. They offer many advantages over traditional financial systems, such as greater security, privacy, and transparency. However, they also face challenges and risks that must be carefully navigated. As the industry continues to evolve and mature, it will be important for investors, businesses, and regulators to stay informed and adapt to the changing landscape of cryptocurrency.

"If your LAZY, I understand...

Just read this instead of the book"

Conclusion Summary

Introduction

Cryptocurrencies are digital assets that use cryptographic technology to secure transactions and control the creation of new units. Unlike traditional currencies, cryptocurrencies are decentralized and operate independently of central banks or governments. This allows for greater security and privacy, as well as greater transparency and control over transactions.

The first cryptocurrency, Bitcoin, was created in 2009 by an individual or group of individuals using the pseudonym Satoshi Nakamoto. Since then, thousands of other cryptocurrencies, also known as altcoins, have been created, each with its own unique features and characteristics.

How Cryptocurrencies Work

Cryptocurrencies work by using a decentralized ledger, or blockchain, to record transactions and control the creation of new units. Each transaction on the blockchain is verified and secured by a network of nodes, or computers, that work together to ensure the integrity and accuracy of the ledger.

To send or receive cryptocurrency, users must have a digital wallet that contains their private keys, which are used to sign transactions and verify ownership of the cryptocurrency. Transactions are typically processed in a matter of minutes, and fees are generally much lower than those associated with traditional banking and financial systems.

Advantages of Cryptocurrencies

There are many advantages to using cryptocurrencies, including:

Greater Security and Privacy: Cryptocurrencies use cryptographic technology to secure transactions and protect user privacy. This makes them more secure and less vulnerable to fraud or theft than traditional financial systems.

Decentralization: Cryptocurrencies are decentralized, meaning that they operate independently of central banks or governments. This allows for greater transparency and control over transactions, and reduces the risk of corruption or manipulation.

Lower Transaction Fees: Cryptocurrency transactions are typically processed at much lower fees than those associated with traditional banking and financial systems.

Global Accessibility: Cryptocurrencies can be accessed and used by anyone, anywhere in the world, provided they have an internet connection and a digital wallet.

Innovation and Disruption: Cryptocurrencies have the potential to disrupt traditional industries and unlock new opportunities for innovation and growth.

Challenges and Risks

Despite their many advantages, cryptocurrencies also face a number of challenges and risks, including:

Volatility: Cryptocurrencies are highly volatile, with prices fluctuating rapidly and dramatically. This makes them a high-risk investment and can lead to significant losses for investors.

Regulatory Uncertainty: Cryptocurrencies are still largely unregulated in many countries, which can create uncertainty and legal challenges for businesses and investors.

Security Risks: Cryptocurrencies are vulnerable to security risks, such as hacking and theft. Users must take careful precautions to protect their digital wallets and private keys.

Lack of Adoption: While cryptocurrencies have gained significant popularity in recent years, they are still not widely adopted or accepted by mainstream businesses and consumers.

Scalability: Cryptocurrencies face challenges with scalability, as the current blockchain technology can only process a limited number of transactions per second. This can lead to slow transaction times and high fees during periods of high demand.

In conclusion, cryptocurrencies are digital assets that use cryptographic technology to secure transactions and control the creation of new units. They offer many advantages, including greater security and privacy, lower transaction fees, and global accessibility. However, they also face a number of challenges and risks, including volatility, regulatory uncertainty, security risks, lack of adoption, and scalability issues.

As the crypto industry continues to evolve and mature, it will be important for businesses and investors to stay informed, conduct thorough research, and carefully evaluate the potential risks and rewards of any investment opportunity. With the right tools and strategies, it is possible to invest in cryptocurrencies safely and securely, and to reap the potential benefits of this exciting technology.

Cryptocurrencies are unique in that they operate independently of central banks or governments. This decentralization allows for greater transparency and control over transactions, as well as reduced risk of corruption or manipulation. Additionally, the use of cryptographic technology makes transactions more secure and less vulnerable to fraud or theft.

The first and most well-known cryptocurrency is Bitcoin, which was created in 2009. Since then, thousands of other cryptocurrencies, or altcoins, have been created, each with their own unique features and characteristics. These altcoins often provide different functionalities or use cases than Bitcoin, such as faster transaction times or increased privacy.

To use cryptocurrencies, users must have a digital wallet that contains their private keys. These keys are used to sign transactions and verify ownership of the cryptocurrency. Transactions are processed through the decentralized ledger, or blockchain, which is maintained by a network of nodes that work together to ensure the integrity and accuracy of the ledger.

One of the major advantages of cryptocurrencies is lower transaction fees compared to traditional banking and financial systems. Additionally, cryptocurrencies are accessible to anyone with an internet connection and a digital wallet, regardless of their location or financial status.

Cryptocurrencies also have the potential to disrupt traditional industries and unlock new opportunities for innovation and growth. For example, decentralized finance (DeFi) projects are using cryptocurrencies and blockchain technology to create new financial instruments and services that are more accessible and transparent than traditional financial systems.

However, cryptocurrencies also face a number of challenges and risks. One of the biggest challenges is volatility, with prices fluctuating rapidly and dramatically. This can lead to significant losses for investors, and makes cryptocurrencies a high-risk investment. Additionally, regulatory uncertainty and security risks are major concerns, as cryptocurrencies are still largely unregulated in many countries and are vulnerable to hacking and theft.

Despite these challenges and risks, the potential benefits of cryptocurrencies have attracted a growing number of investors and businesses. As the industry continues to evolve and mature, it will be important for stakeholders to navigate these challenges and risks with caution, while also embracing the potential for innovation and growth that cryptocurrencies offer.

One area of innovation that has emerged in recent years is decentralized finance (DeFi). DeFi refers to a range of financial applications that use blockchain technology to create new financial instruments and services that are more accessible and transparent than traditional financial systems.

DeFi projects often rely on cryptocurrencies to operate, and have been growing rapidly in recent years. One example of a DeFi project is decentralized exchanges (DEXs), which allow users to trade cryptocurrencies without the need for a centralized exchange. This can provide greater transparency and security, as well as lower fees compared to traditional exchanges.

Another area of innovation is non-fungible tokens (NFTs), which are unique digital assets that are verified on a blockchain. NFTs can be used to represent a variety of assets, such as art, music, and collectibles. They have become increasingly popular in recent years, with some NFTs selling for millions of dollars.

As the crypto industry continues to mature, it is likely that we will see new and innovative use cases emerge, as well as further developments in blockchain technology. However, it will also be important for regulators and stakeholders to address the challenges and risks associated with cryptocurrencies, such as volatility, security, and regulatory uncertainty.

Overall, cryptocurrencies are a promising and exciting development in the world of finance and technology. They offer many advantages over traditional financial systems, such as greater security, privacy, and transparency. However, they also face challenges and risks that must be carefully navigated. As the industry continues to evolve and mature, it will be important for investors, businesses, and regulators to stay informed and adapt to the changing landscape of cryptocurrency.

Chapter 1.

"The Origins of Cryptocurrencies: A Brief History of Bitcoin and Altcoins": This chapter provides a brief history of cryptocurrencies, focusing on the creation of Bitcoin and the subsequent emergence of altcoins.

Chapter 2.

"Understanding Blockchain Technology: How it Powers the Crypto Revolution": This chapter explores the underlying technology that powers cryptocurrencies, known as blockchain, and how it has revolutionized the way we think about digital transactions and data storage.

Chapter 3.

"Bitcoin: The Original Cryptocurrency and Its Impact on the World of Finance": This chapter delves into the impact that Bitcoin has had on the world of finance, including its potential to disrupt traditional financial systems and its use as a store of value and investment.

Chapter 4.

"The Pros and Cons of Investing in Cryptocurrencies: A Comprehensive Analysis": This chapter provides a comprehensive analysis of the pros and cons of investing in cryptocurrencies, including factors such as volatility, regulatory uncertainty, and security risks.

Chapter 5.

"Crypto Regulations Around the World: How Governments are Responding to the Rise of Cryptocurrencies": This chapter examines how governments around the world are responding to the rise of cryptocurrencies, including the challenges of regulating a decentralized and borderless technology.

Chapter 6.

"Altcoins: Exploring the Diversity of Cryptocurrencies Beyond Bitcoin": This chapter explores the diverse range of altcoins that have emerged in recent years, each with their own unique features and characteristics.

Chapter 7.

"The Future of Crypto: Predictions and Trends for the Next Decade": This chapter looks to the future of cryptocurrencies, examining predictions and trends for the next decade, including the potential for greater adoption and mainstream integration.

Chapter 8.

"DeFi: The Emergence of Decentralized Finance and its Relationship to Cryptocurrencies": This chapter explores the emergence of decentralized finance (DeFi) and its relationship to cryptocurrencies, including how DeFi projects are using blockchain technology to create new financial instruments and services.

Chapter 9.

"Crypto Security: Best Practices for Keeping Your Digital Assets Safe": This chapter provides best practices for keeping your digital assets safe, including tips on securing your digital wallet and protecting your private keys.

Chapter 10.

"The Psychology of Crypto: Understanding the Emotions and Behaviors of Crypto Investors": This chapter examines the emotions and behaviors of crypto investors, including the impact of FOMO (fear of missing out), the influence of social media and online communities, and the importance of emotional discipline when investing in cryptocurrencies.

Overall, these 10 chapters provide a comprehensive overview of the world of cryptocurrencies, exploring their origins, technology, impact on finance, and future potential, as well as examining the challenges and risks associated with investing in this emerging and rapidly evolving field.